How the Second Grade Got $8,205.50 to Visit the Statue of Liberty

Nathan Zimelman

Illustrated by **Bill Slavin**

Harcourt Brace & Company

Orlando Atlanta Austin Boston San Francisco Chicago Dallas New York Toronto London

Report on the drive to collect funds for a visit to the Statue of Liberty. Susan Olson, treasurer and reporter, second grade, Newton Barnaby School.

Old Newspapers, Cardboard Boxes, and All-Kinds-of-Paper Drive

The second grade collected two tons of all kinds of paper for which we were paid thirty dollars by Mr. Abner Carmody, recycler of everything.

Expenses:

Two dollars which we had to pay five younger brothers and sisters to borrow their wagons which had been our wagons in the first place.

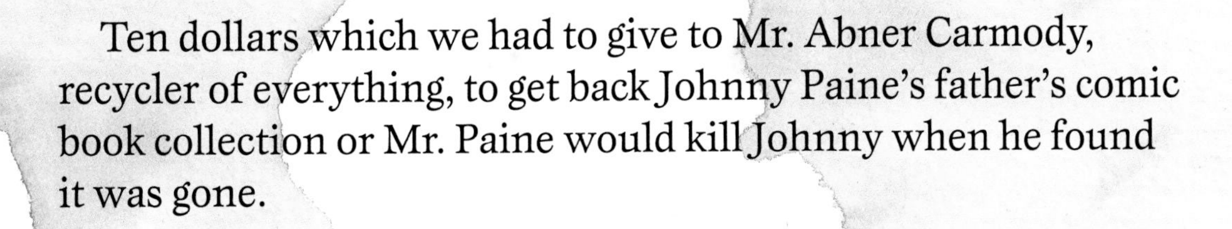

Ten dollars which we had to give to Mr. Abner Carmody, recycler of everything, to get back Johnny Paine's father's comic book collection or Mr. Paine would kill Johnny when he found it was gone.

Five dollars which was the cost of the ticket our principal, Mr. Harold Hope, got because Robert Rose, who was sent with a quarter to put in the parking meter, got worried about a dog fight and forgot. (Mr. Hope's regular parking space was covered over by two tons of old newspapers, cardboard boxes, and all kinds of paper.)

Profit:

The old newspaper, cardboard boxes, and all-kinds-of-paper drive has made a profit of thirteen dollars.

We are practically on our way.

Lemonade Stand

Expenses:

No charge for thirty lemons which Cecil Cooper's neighbor's tree contributed from the branches that hang over into Cecil Cooper's yard.

No charge for ten pounds of sugar which was given by Geraldine Arthur's mother who was starting her weekly diet.

No charge for a washtub which was loaned to us by Mr. Abner Carmody, recycler of everything, for a free drink of lemonade whenever he passed by.

15

No charge for glasses, a small table, a dipper, water, and ice cubes which we got from Eleanor Robins's mother, who is always giving us stuff so Eleanor will be popular.

It helps.

In total the expenses for the lemonade stand were no dollars and no cents.

Profit:

Our income from the lemonade stand was no dollars and no cents because after we had set up the tub under the shady elm tree and filled it with water and ice cubes and squeezed the lemons and mixed in the sugar, Eleanor's cat fell out of the tree into the tub, and Eleanor's mother made us pour out all the lemonade—after we fished out the cat.

However, I am happy to report that the lemonade stand took in ten dollars from Eleanor Robins's neighbor, Mr. Norbert Norvill. He was laughing when he gave us the money.

I don't think he likes Eleanor's cat.

Baby-Sitting and Dog Walking

Expenses and Profit:

Nothing and nothing. Nobody trusts a second grader to baby-sit anybody, even baby brothers and sisters.

All the dogs in this town walk each other.

Candy Sales

Expenses:

No charge for three pounds of chocolate fudge full of walnuts which was hurried out the back door and handed to us by Scott Alexander's mother after he brought a five-cavities report home from the dentist.

One dollar for one box of aluminum foil to fancy wrap the candy for sale.

Refund of one dollar for one box of aluminum foil which was not needed after Geraldine Arthur's mother ran out of her house screaming, "Tomorrow starts another week to start another weekly diet!" and bought the three pounds of unwrapped chocolate fudge full of walnuts.

Profit:

Our candy sale profit was seven and a half dollars which Geraldine Arthur's mother gave us after her first beaming bite.

We are getting closer to the Statue of Liberty!

Car Wash

Expenses:

No charge for three boxes of genuine non-detergent, non-anything, pre–World War II soap flakes traded by Mr. Abner Carmody, recycler of everything, for three car washes to be given as soon as he finishes putting together his car from parts which he has been saving.

SOAP

GAS

Five dollars which we paid to Mr. William W. Williams for the use of the empty service island where he is going to put in a second gas pump as soon as business gets better.

No charge for sponges, towels, and buckets contributed by parents when we promised to remember to bring everything back.

Because of all the excitement we kind of forgot to pick them up and somebody else kind of went off with them. The sponges, towels, and buckets will cost an extra twenty dollars to replace.

That makes twenty-five dollars worth of expenses.

Profit:

After Charlie Hendon pretended to throw a bucket of water through the open window of the first customer, our principal, Mr. Harold Hope, and the water kind of got away from Charlie, Ms. Pinckney, our teacher, and Mr. Hope had to go around the corner to get a cup of coffee to settle their nerves.

So they were not there when Charlie began showing Cecil Cooper what his father had taught him about driving and accidentally released the brakes and sent Mr. Hope's car rolling backward into the street, aimed right at the doors of the First National Bank.

Ms. Pinckney and Mr. Hope were also not there when the doors of the bank burst open, and two men waving guns and carrying canvas sacks rushed out.

When they noticed Mr. Harold Hope's car rolling right at them…

the men dropped everything and jumped back through the doors of the bank where they were captured by a smiling Ms. Henrietta Vinker, the First National Bank's friendly guard.

The second grade will receive a reward of eight thousand, two hundred dollars.

Since the high curb in front of the bank bumped Mr. Hope's car to a stop without any damage, the total profit from the car wash was eight thousand, one hundred seventy-five dollars.

Total Profit on Everything:

Eight thousand, two hundred five dollars and fifty cents.

This will be way more than enough to pay for our visit to the Statue of Liberty, but Chief of Police Gates says that they take forever and a day to pay reward money, and we will probably be the third grade before we see a penny of it.

This will not delay the visit! The parents of the second grade got together and collected enough money from each other for the trip.

They said they needed the rest.

WE ARE ON OUR WAY.

For Sarah Zimelman.
N.Z.

For Mr. Ernst Hadorn,
recycler of everything.
B.S.

This edition is published by special arrangement with Albert Whitman & Company.

Grateful acknowledgment is made to Albert Whitman & Company for permission to reprint *How the Second Grade Got $8,205.50 to Visit the Statue of Liberty* by Nathan Zimelman, illustrated by Bill Slavin.

Printed in the United States of America

ISBN 0-15-309680-2

2 3 4 5 6 7 8 9 10 035 2000 99 98